Dear Parents,

Here at edZOOcation, it's our mission to inspire a new generation of Wildlife Guardians. We support zoos, conservation efforts, and animal education organizations all over the world. Just as importantly, we want to help foster your child's love of animals.

Our books are filled with amazing facts and stunning photographs designed to bring science to life. We follow Common Core standards and state science standards to teach your child about nature at a level that's just right for them.

Nurture your child's love of learning and love of animals by reading with them today!

Sincerely,
Jenny Curtis, Founder

Dedication:

For London, who's almost as fast as a cheetah and at least twice as wild.
—S.K.

For Wyatt, and his running adventures.
—A.R.

Copyright © 2025 Wildlife Tree, LLC. All rights reserved.

Note: This is a reprint with an updated cover design in 2024.

Author: Sara Karnoscak

Designer: Allyson Randa

Editor: Tess Riley

Photo Credits:

AdobeStock.com

Pixabay.com

Pexels.com

ISBN: 978-1-965081-11-2

This book meets Common Core and Next Generation Science Standards.

Table of Contents

4	A Cheetah's Body
8	Tracks and Poop
10	Species
14	Families
16	Talk
17	Where Cheetahs Live
21	Sleeping and Eating
24	A Day in the Life of a Cheetah
26	The Food Web and Dangers
28	Cheetah Fun
30	Glossary
32	Silly Cheetahs

RUN!

Cheetahs are the fastest runners. Prey tries to dodge them. But cheetahs can turn fast as they run. They can't run fast for long, though. After they run fast, they need to catch their breath.

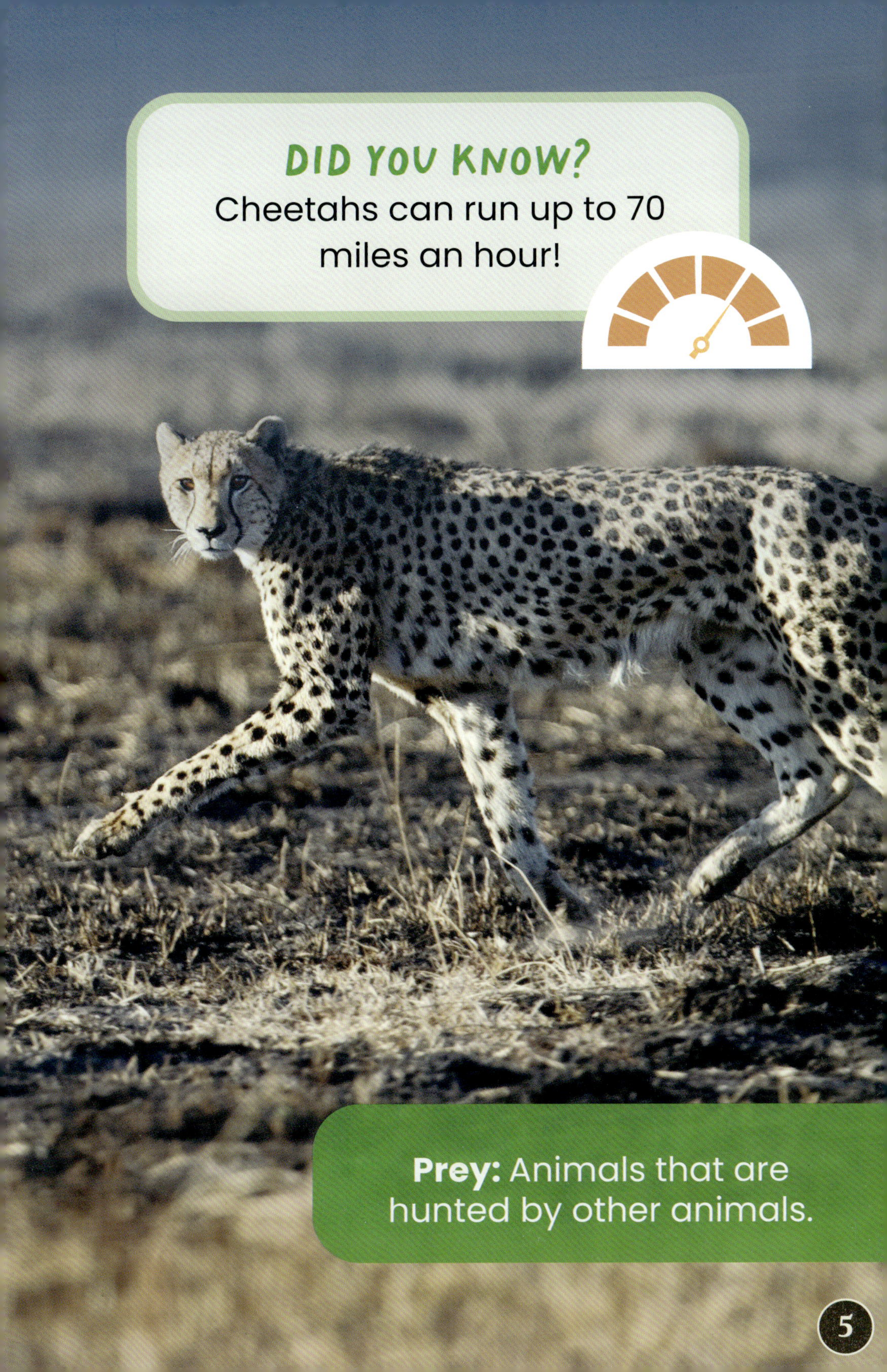

DID YOU KNOW?
Cheetahs can run up to 70 miles an hour!

Prey: Animals that are hunted by other animals.

A CHEETAH'S BODY

Big chest and lungs.

Long tail to steer.

Tear marks to keep sun out of their eyes.

Claws and feet that grip the ground.

Cones on front legs for brakes.

CHEETAH TRACKS

Scientists use cheetah tracks to help them. Cheetah tracks show them where cheetahs are. It also helps them know how big they are.

Prey: Animals that are hunted by other animals.

CHEETAH POOP

Scientists use cheetah poop to help them, too. Cheetah poop helps them know what cheetahs eat. It also helps them know which cheetahs are family.

Big and Small Cats

Cheetahs are smaller than lions and tigers. Cheetahs are much bigger than house cats.

SPOT THE CHEETAH.

Cheetah spots are black dots.

Leopard spots are like roses.

Jaguar spots have spots.

CHEETAH CUBS.

Mother cheetahs have two or more cubs at a time. Cubs are helpless when they are born. They are born with a mane. The mane helps them blend in with grass.

Cubs: Baby cheetahs.

Mane: Long hair on an animal's neck.

CAMOUFLAGE

Blending in helps them hide from other animals. When they are older, their mother teaches them to hunt.

CHIRPING CHEETAHS

Cheetahs can't roar! Cheetahs can make other sounds like...

PURR

CHIRP

BARK

GROWL

Can you sound like a cheetah?

CHEETAHS AT HOME

Cheetahs live in grasslands and deserts. They live where there is food to catch. Female cheetahs live alone. Male cheetahs live with their brothers.

Grasslands: Big, open land covered in grass.

THE LAST CHEETAHS

Long ago, there were very big cheetahs, and there were cheetahs around half the world. There are roughly 7,000 cheetahs left.

WHERE ARE THEY?

Can you spot the cheetahs in these pictures?

WHERE DO CHEETAHS LIVE?

Cheetahs used to live in many places. Now cheetahs live in Africa. A few cheetahs still live in Iran.

CAT NAPS

Most big cats are awake at night. Cheetahs are awake in the daytime. They like to hunt when they can see their prey well.

WHO IS IT?
Can you tell which big cat is a cheetah?

On the Hunt

CHEETAHS SIT ON HILLS OF DIRT TO WATCH FOR PREY.

THEN THEY SNEAK UP ON THEIR PREY.

THEIR COATS BLEND IN WITH THE GRASS.

THEN THEY'RE OFF!

THEIR SPEED MAKES THEM VERY GOOD HUNTERS.

23

A Day In The Life of A Cheetah

THE CHEETAH WAKES UP IN THE MORNING.

IT'S NOT TOO HOT YET. IT'S THE PERFECT TIME TO HUNT.

THE CHEETAH AND HIS BROTHERS HUNT.

THEY CATCH A GAZELLE.

AFTER THEY EAT, THEY REST.

THEY HUNT AGAIN AT THE END OF THE DAY.

NOW TIME FOR BED!

THE FOOD WEB

A food web shows how living things need each other for food. Cheetahs hunt animals for food. Other animals hunt cheetahs for food.

DANGERS

There are not many cheetahs left. Some people kill cheetahs to protect their farms. Dogs that scare cheetahs away from farms can help cheetahs.

SILLY CHEETAHS

Cheetah, cheetah, fast and strong,
Running, running all day long.
Through the grass so tall and wide,
With a spotted coat to hide.

Q: Why did the big cat get in trouble at school?
A: Because he was a cheetah.

Q: Why did the cheetah always lose at hide and seek?
A: Because he was spotted.

Q: What time is it when a cheetah takes your backpack?
A: Time to get a new backpack.

Q: What do you get when you mix a cheetah and a rhino?
A: A Cheeto

Q: What do you get when you mix a cheetah and a burger?
A: Fast food.

Q: Why do cheetahs eat raw meat?
A: Because they don't know how to cook.